PLEASE WASH
YOUR HANDS
BEFORE YOU READ ME
AND KEEP ME CLEAN

For Judy, my sister.

This book belongs to:

_____

written, designed, and illustrated by

Jan Adkins

HEAVY EQUIPMENT

Charles Scribner's Sons New York

Library of Congress Cataloging in Publication Data
Adkins, Jan. Heavy equipment.
SUMMARY: Detailed drawings and brief text document the shape and function of such heavy machinery as compactors, excavators, graders, crawlers, and cranes. 1. Construction equipment—Juvenile literature. 2. Road machinery—Juvenile literature. [1. Construction equipment. 2. Road machinery] I. Title. TH900.A35   624'.028
80-15213   ISBN 0-684-16641-0

1 3 5 7 9 11 13 15 17 19 Q/D 20 18 16 14 12 10 8 6 4 2
Printed in the United States of America

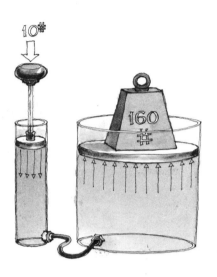

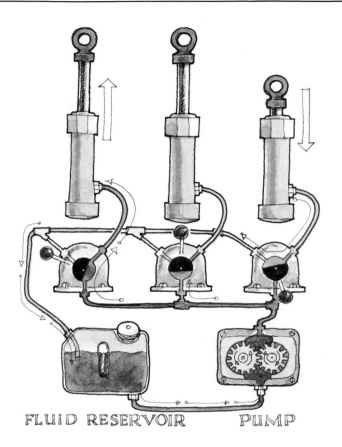

FLUID RESERVOIR   PUMP

They are the big machines, the heavy equipment that chews at the earth and builds on it. They are so strong! The power of a thousand horses lives in their metal hearts and nothing can stand against them. They carve down mountains and lift valleys up.

One reason for the power of the big machines is the hydraulic (hi *drawl* ik) cylinder. A fluid, like water or oil, carries pressure evenly. Pressure from a small piston is carried through the fluid to a large piston. In the head of the small piston fluid is pressing on a small area, but

HYDRAULICS

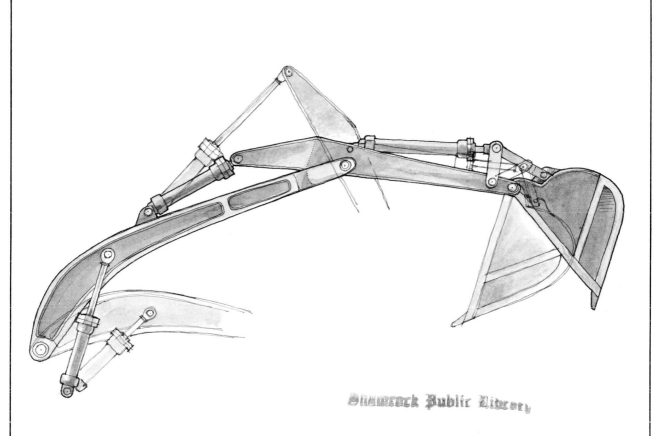

when the fluid presses with equal force on every square inch of the large piston head a greater force is applied.

In the workings of big machines a pump puts special oil (from the fluid reservoir) under pressure. The operator of the machine opens a valve that sends oil into a piston. The piston expands, with enormous force, and lifts an arm or swings a bucket. When the valve is reversed oil is released from the piston — quickly or slowly as the operator wishes — and it returns to the reservoir.

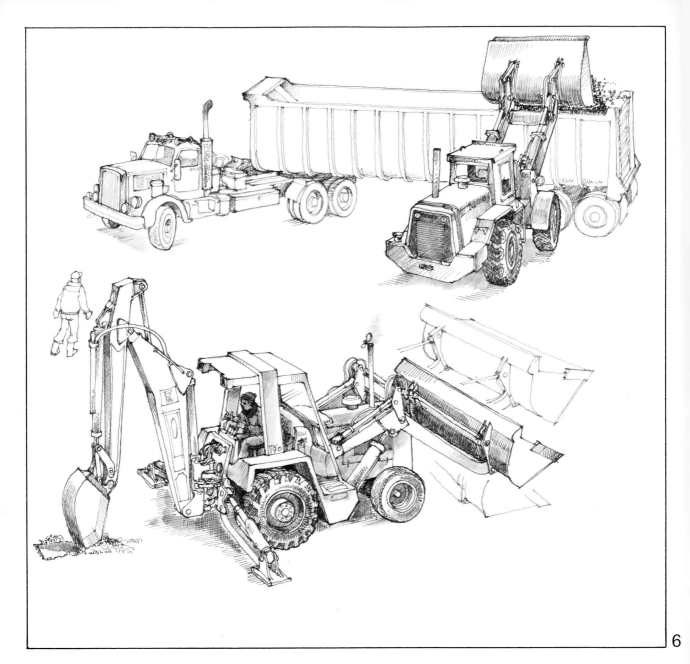

BACKHOE/LOADER

Bucyrus-Erie 190 Series III Backhoe/Loader
Terex 72-51B Loader

You can see a backhoe/loader at almost any building site. They are the little brothers of the big machines, busy at digging out ditches and foundations with one end and dealing with piles of dirt with the other end. A *front end loader* picks up piles and dumps them into trucks to be taken away.

A grader smooths and levels and makes an even surface. This grader has *scarifiers* in front of the blade; they are hooks that tear up and loosen hard soil.

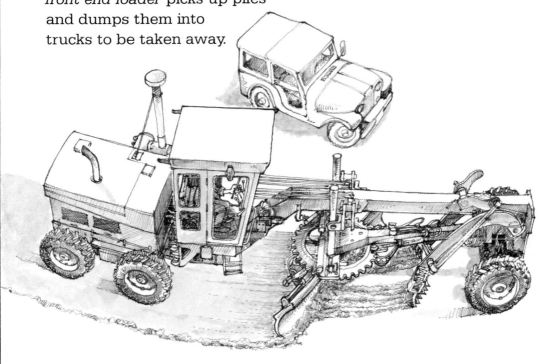

7

GRADER

Champion Articulated 740 A Motor Grader

Compactors are finishing tools that pack soil into a drive-able surface or smooth and press down a new asphalt road. The wicked studded steel tires of the *landfill compactor* tear trash to bits and crush it down to put away more garbage in every square foot.

8

COMPACTORS

Ray Go Rascal 400-A Vibratory Soil Compactor
Ray Go Rustler 6604-A Double Drum Asphalt Compactor
Ray Go Ram-Pak 65 Sanitary Landfill Compactor

Not all heavy equipment moves earth. A few just move heavy things. Small forklifts scuttle about factories and warehouses. This one is fueled by a tank of natural gas. The big forklift lifts and stacks cargo containers on the docks.

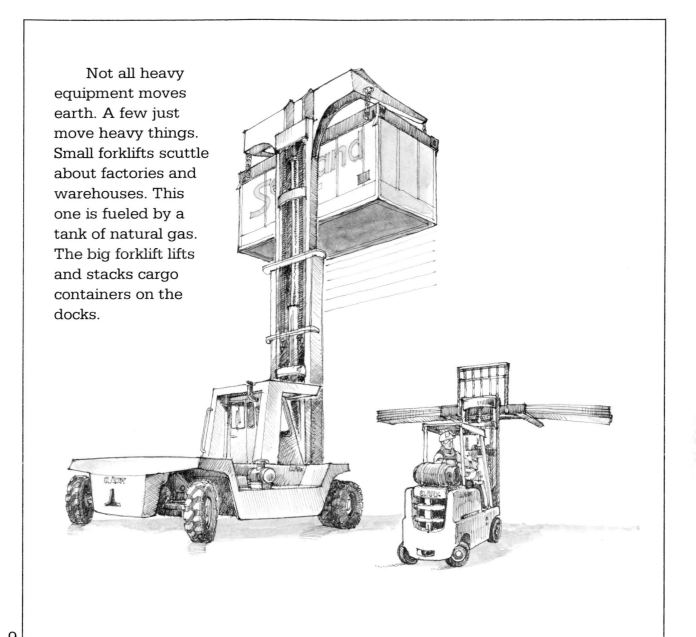

9

Clark 300-40 Cushion Tire Forklift
Clark 500 Y 800 D Lift

FORKLIFTS

Like clumsy mechanical horses, crawlers snort and rumble across the landscape, pushing away anything in their path. Their broad tracks and track spurs give them the traction to lump up earth and rock. Some are small. Some are large. Some, like the Caterpillar D 10, are almost terrifying. They root up the landscape with their blades and the rippers behind them, then they reshape it. Maybe they are a little more like mechanical pigs.

CRAWLERS

11

Terex 82-80
Caterpillar D 3
Caterpillar D 10

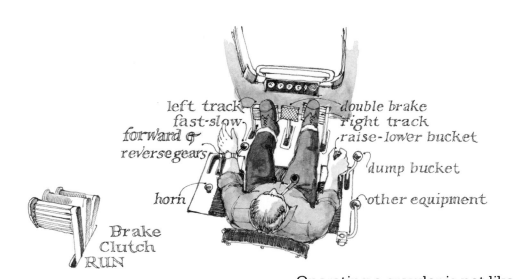

left track
fast-slow
forward &
reverse gears

double brake
right track
raise-lower bucket

dump bucket

horn

other equipment

Brake
Clutch
RUN

Operating a crawler is not like driving a truck. The right and left tracks are controlled separately by levers or foot pedals. To turn, the operator releases power from one track or stops it entirely; the other track continues and pushes the crawler around. On some crawlers one track can be reversed while the other travels forward and the crawler spins around in its own length. A good operator works a crawler delicately with his hands and feet, like a piano player.

CRAWLER

Caterpillar 983B Crawler Loader

CRAWLER

Caterpillar D5 LGP

With special tracks, longer and wider, a crawler can work in wet, marshy ground where a wheeled machine would sink. The area that presses down — even on mud — is large enough to support its weight in the same way that a snowshoe supports a person's weight on delicate snow when a boot would sink in.

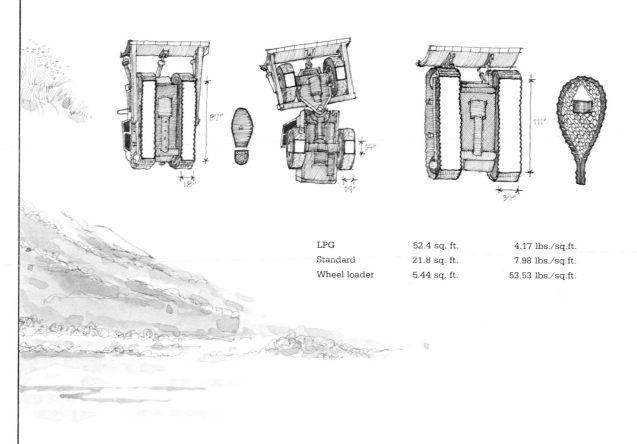

LPG	52.4 sq. ft.	4.17 lbs./sq.ft.
Standard	21.8 sq. ft.	7.98 lbs./sq.ft.
Wheel loader	5.44 sq. ft.	53.53 lbs./sq.ft.

LOW GROUND PRESSURE

It's a mobile arm with a long, hungry reach. Its boom, stick, and bucket dig deep with surprising delicacy. Its operator can make it do everything but waltz.

stick & swing

engine controls & instruments

tracks

boom & bucket

BOOM

STICK

BUCKET

EXCAVATOR

17

Bucyrus-Erie 350 H Hydraulic Excavator

18

LOGGERS & SKIDDERS

You want tough machines to bring lumber down from the timberland, carting and skidding two-ton logs over improvised roads. Skidders, with tracks or wheels, rough out their own roads and use their winches to snake up logs for dragging. A logger's pincers pluck up logs to load them onto timber-hauling trucks.

Terex 82-20 B Crawler
Caterpillar 518 Skidder
Terex 72-61 Logger

Big weights on one side balance the heft of a pipe length on the other. With its winch, A-frame, and treads, the pipelayer puts its load just where it wants to.

20

PIPELAYERS

Caterpillar 594 H Pipelayer

They are like loud insects. They crawl over quarries and over the rock ledges that crop up in the way of roads and buildings. Their air compressors make their power.

They drill their holes and move away while explosive charges are gently lowered into the holes. Suddenly the rock is more manageable.

Worthington Wrangler Crawler drills

CRAWLER DRILLS

Scrapers lower their *bowls* and open their *aprons* to take a cut of earth six to twenty-four inches deep. They close the apron, raise the bowl, and rumble off with the earth. An *ejector* pushes the earth out. In difficult soil a crawler sometimes pushes a scraper for extra force. Some scrapers are powered on both ends. The conveyer in an *elevating scraper* lifts up the soil and drops it into the bowl.

SCRAPERS

Caterpillar 621 B Wheel Tractor-Scraper
Caterpillar 623 B Elevating Scraper
Caterpillar 666 B Tandem Powered Wheel Tractor-Scraper

OFF-ROAD TRUCKS

Highways are not big enough for them. They would take up two or three lanes. They travel roads built for them and they travel fast, with a trail of dust and gravel thrown up behind them. They carry ore, *overburden* (the earth that lies over coal or ore), rock, and debris. The earth shakes when they pass.

Caterpillar 773 Off-Highway Truck, 50-ton capacity
The Terex Titan, Model 33-19, 350-ton capacity

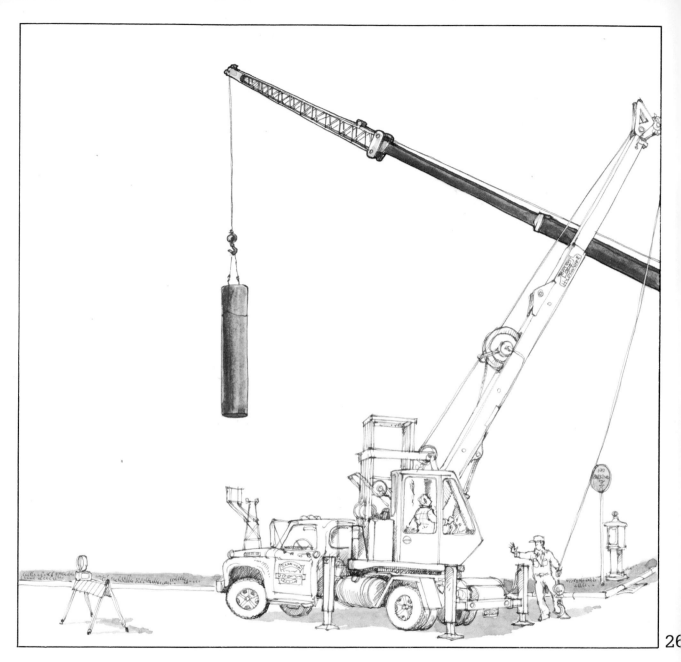

TRUCK CRANES

They travel to the site on their own power. They drive into postion and set up their outriggers to make a wide base. They rig their steel cables and extend their long, telescoping arms. A rigger on the ground hooks up a load and moves back to talk with his hands and arms, helping the crane operator judge the hoists, drops, angles, and distances. Together they play a dainty, dangerous game of blocks and jackstraws.

7

Bucyrus-Erie H-3 5-ton Hydrocrane
Bucyrus-Erie 60-XC 60-ton Hydraulic Truck Crane

The same cab, crawlers, and machinery, rigged differently, make a shovel and a dragline crane. The shovel scoops from the front and dumps its load out the bottom of the bucket into trucks or railroad cars. The crane with a dragline rig reaches out and winches its bucket back, a powerful and efficient way to move earth, dig underwater channels, and load gravels and ores. The same crane fitted with different buckets can dig with a *clamshell,* lift metal with an *electromagnet,* clean out catchbasins with a *spring bucket,* or pick up hay, lumber, and bales of scrap with a tined grapple.

CRANES

Bucyrus-Erie 88-B Shovel and 88-B Dragline

9

SHOVEL & DRAGLINE

Big Muskie is the largest brother in a family of giants. For now it is the largest land machine that has ever moved on the earth. Its masts rise over the hills of Ohio and its lights at night can be seen in the next county. Its job is to strip off the earth that covers layers of coal. Each pass of its dragline bucket engulfs 220 cubic yards of shale and soil, about 320 tons. The bucket can comfortably hold eighteen Jeep CJ-5's. To move,

GEM

Big Muskie slides two jointed shoes (130 feet long) about fourteen feet forward; with hydraulic cylinders the shoes take on most of the rig's 27 million pounds and drag the flatbottomed tub under the housing along. It moves about 900 feet in an hour. Altogether it is 487 feet long, 151 feet wide, and its boom reaches up 222 feet. It is awesome.

GIANT EARTH MOVER

The big machines can clear off a forest for the lumber, lift off the soil to bring up ore and coal, fill a swamp to let a road pass. But in all the ripping, tearing, and filling they can leave a dead land. Are they good or bad?

They are not good and not bad. They are like a great rain or a deep winter that can do harm and good. The big machines are so strong that they can be dangerous as well as wonderful...it depends on the men and women who use them. If the people who use them are quick and greedy, the big machines will destroy. If the people are patient and thoughtful, the machines will build. Heavy equipment has no mind of its own.

Many kind and patient people have given time and attention to this small book. Thanks to them and to their companies: Bucyrus-Erie; Caterpillar; Terex, a division of General Motors; Jeep, a division of American Motors; Champion; Ray-Go; Clark; Worthington.

These drawings were made on single weight illustration board with technical pens loaded with Pelikan plastics ink and with dilute ink; ivory black washes were laid over with sable brush.

The body type is Serifa 55, the display is Granby, both set by Arrow Typographers. The paper stock is 80-pound matte-coated Frostbrite. Binding by A. Horowitz & Sons, Bookbinders. Printing by Halliday Lithograph.